I0486724

Rise of Automation - Technology and Robots Will Replace Humans

Adidas Wilson

Published by Adidas Wilson, 2017.

While every precaution has been taken in the preparation of this book, the publisher assumes no responsibility for errors or omissions, or for damages resulting from the use of the information contained herein.

RISE OF AUTOMATION - TECHNOLOGY AND ROBOTS WILL REPLACE HUMANS

First edition. December 3, 2017.

Copyright © 2017 Adidas Wilson.

ISBN: 978-1393907022

Written by Adidas Wilson.

---

1.   http://www.adidaswilson.com

Table of Contents

Introduction

You probably have an idea how robots will affect human workers negatively. Chief players in the tech world like Bill Gates and Elon Musk have provided their solutions; universal basic income or robot tax. But amidst the serious warnings and the utter sci-fi utopias, the human pain that will follow future job loss seems to be forgotten. 15 years or so from now, the US economy will lose 38% of its jobs to automation. This rate is alarming. And yet, many people maintain that automation should not and cannot slow down.

However, what if the progress is decelerated a little? Just enough to match the slow fashion and slow food trends maybe? At the very least, people should rethink the ownership of autonomous trucks. Robotization would not be that bad if truck drivers owned the automatic trucks instead of having a corporation own them all. In the meantime; robotization is a real threat and poses a danger to crucial human infrastructure. Maybe you should understand what human infrastructure is first. Infrastructure refers to fiber optic cables, roads, power plants, and electricity grids among others. Human infrastructure is a term that shows us that people are also "essential servicers." Critical human infrastructures are the people that are most threatened by robotization — like truck drivers. If autonomous vehicles are introduced, they will barely get employed elsewhere.

Instead of trying to make robots more human-like, the primary focus should be on human workers, even if it means coming up with a cooperated approach for ownership of the autonomous trucks—at least then, most drivers will not be left jobless. Other jobs like (cashiers, movie ushers, legal assistants); maybe legislation should be developed to help all classes of workers who could be displaced by robots. All this means one thing: the process of automation must be slowed down.

Again, autonomous trucks are very easy to hack, planning now will prevent (or slow down) a major worldwide security threat. Imagine what hackers can do with automated trucks. Some organizations have already voiced their concerns and are completely against automation. How will truck drivers feed their families? Uber has promised automation very soon, yet some drivers took loans to buy the cars for Uber. How would they pay for loans when automation comes, no one knows. Over the past couple of years, researchers have seen the risk of robots taking away jobs from humans. Up to ten million jobs or more will disappear as robots keep advancing, especially the unskilled ones. Some experts maintain there is still nothing to worry about, holding on to the perception that technology will create more jobs and replace the old ones. This has always been the case, but the stakes are high, and white collar jobs are among the ones at risk. So far, automation has claimed five white collar jobs.

Creating information that can captivate others, attract them to open an email, and make a purchase might sound like an ability endowed to humans only. But this is not the case, software that can carry out natural language by analyzing data using semantic algorithms are standing up to the stage. The software can thus determine ideal subject lines for emails and are capable of taking over the once human-dominated online marketing. Another area where machines are taking the lead is the programmatic ad-buying sector. The software can check online, use vast information on potential subjects to target ads on ideal prospects, all in an instance.

The other job that is apparently on the verge of toppling is the duty of financial advisors and analysts. The biggest threat is that computing power, combined with predictive systems and big data are providing the needed analysis and predictions that investors need. This means some financial professionals might have a difficult time hanging on as machines flood the field. If there is one sector of humanity that involves heavy use of documents, then it is the world of law. Reviewing these extensive materials was once the work of paralegals and lawyers at the lower level, but software is doing the job quite well nowadays. Besides, machines are capable of carrying out other duties involved in the legal world. This means E-discovery, law firm associates, and others in the industry are already facing stiff competition from the AI front. Gathering the information and presenting it to the audience is all in a reporter's day. Factor in the aspect of readability on the machine and the risk of disruption becomes inevitable. Some media like Associated Press are experimenting with using software that can create reports on corporate earnings automatically. The process has advanced over time to become error-free, which means it can outperform a human worker, putting the job of a reporter in jeopardy.

While doctors might be considered the most appropriate hands-on experts when it comes to treatment, the tables are turning on diagnosticians, Anesthesiology, and surgeons. Some machines have been FDA approved to carry out low-level anesthesia in such settings as colonoscopies. Besides, machines can carry out other tasks as well, while one doctor can work with several robots to control human elements. These machines are also handy in surgery and diagnosis, which means we might soon have the tools working on patients rather than doctors doing the job. It is apparent that robotics is here to stay, and although the fear of these machines taking over jobs becomes real, the reality is inevitable. It is also crucial that you become indispensable in the workplace, which can give you an edge over the competition. Above all, critical thinking comes in handy when it comes to considering yourself successful in the labor world. Critical thinking involves building concepts skillfully and actively and analyzing them, synthesizing them as well as using them for better performance. This skill is crucial to your success in work and an important asset that employers need. Gathering information and making reliable decisions is part of this aspect and helps select the most applicable solution to problems after sorting through numerous solution options.

You might have the talent every employer is looking for or the skills set for the job. But if you are not adaptable to changes in the workplace, then you might be headed for a major failure. The world is ever changing, and this will keep on rolling. The best aspect you can have to succeed anywhere, whether in the workplace or beyond, is being ready and able to adapt to changes with ease. Customer demands will keep changing, and such, you need to be able to identify the shifts in the trends and be able to foresee trends and keep in touch with what is happening around you. In essence, being ready for the future of work is more or less all about you. Remember to equip yourself with creativity and intelligent skills as well.

Chapter 1

Elon Musk and Universal Basic Income

During the World Government Summit in Dubai, Elon Musk brought up crucial ideas on the future of humanity. According to Musk, Universal Basic Income (an economic idea that suggests everyone receives a paycheck from the government for personal spending) is among the few solutions for robotic automation.

When automation becomes widespread, everything will change. People are referring to it as the coming of the "post-scarcity economy." In simple terms, in future (a future that is fast approaching), money will not be a big deal, and all economies will completely collapse. Post-scarcity is one of those things that everyone should try to understand. Traditional economies are still functional because things are hard to come by. For instance, food is limited; otherwise, it would be free. You cannot possibly charge for what is infinite like the sun, can you?

With replicators, which are technically magical boxes that can create anything out of anything in the blink of an eye, things do not have inherent value. You cannot even influence the demand and supply of anything since the demand is not definite and the supply is limitless. In a system like that, the traditional economy is useless and powerless. How would you even draw the demand and supply curve for such a circumstance? Humans are nowhere near replicators, obviously. And a total post-scarcity system will not be possible for a long time. However, great advancements are about to happen. The number of jobs that robots will grab from humans is significant. Take self-driving cars, for instance, that alone will render about 20% of employees unemployed.

That is hardly the only industry that will be shaken. So many jobs can be automated easily and eventually causing millions of Americans to become jobless. It might take twenty years—or even thirty—but the truth is that no economy can withstand that. Plans for universal basic income suggest tax robots. This aims at replacing the revenue that will be lost from the unemployed and alternatively created from the robots. Businesses will still gain—money will keep coming in for people to continue spending. According to some conservative estimates, the robots might even pay for themselves several times over. So companies have nothing to lose.

At the moment (and if nothing changes), this is the only productive solution. Most markets and corporations will remain intact while working with the complicated reality of robots for labor. So somehow, Musk is right. Take time and learn, know the possibilities of the future—because no matter what your job is, it is not safe.

Chapter 2
Silicon Valley and the Automated Future

For quite some time, Antonio Garcia Martinez was living his dream as a Techi in Silicon Valley; until a few years ago. He had just sold his ad company to Twitter and was a senior executive at Facebook. But in 2015 when he thought about the future and realized what a pale world it would be. It was nothing like the glamorous utopia that his colleagues were promising. Antonio's biggest worry about modern technology is the risk that the combined forces of artificial intelligence and automation pose for the world. The development is taking place faster than people realize and another industrial revolution is on the verge of happening. It might destroy the economy and lead to the loss of jobs.

In about 30 years, most people will be jobless—and things will not be good. And if this does not sound serious to you, you should know that Antonio has already prepared a getaway just in case things go south. And he is not the only one who foresees the economic losses. Co-founder of LinkedIn, Reid Hoffman, confirmed that many of the Silicon Valley billionaires already have some form of "Apocalypse insurance." Others have created secret Facebook groups to talk about survivalist tactics. Many tech entrepreneurs are pessimistic about the future—the same one they are helping to create. They say that it's inevitable because technology is unstoppable; like a force of nature. A few years ago, this would have been dismissed as American paranoia for dystopia. However, the robot apocalypse is becoming a potential reality every day. Giant financial publications have been publishing articles implying that the economy is just about to experience a major transformation. Obama's team even suggested that very soon, automated cars will send over 3 million drivers home.

Contrary to what Hollywood portrays, the A.I. revolution will not come in a glossy form. It comes in the shape of smart machines that rely on algorithms, and they will keep getting smarter. Some of these robotic devices are already being used in small scale, but every year more will be introduced. Every industry will be affected. The new type of machines being introduced are those that can solve problems on their own without relying on people to feed them examples, some neural network chatbots were introduced to Facebook recently—but they went rogue, inventing their own language and they had to be shut down.

A.I is not entirely bad. It will significantly improve the medical field and save a lot of lives, and even autonomous trucks will be safer than tired human drivers. Lucrative tech jobs will come up; however, most of the jobs will cease to exist. On the bright side, jobs that cannot be replaced by automation may rise—a few years ago nobody would have thought that app development would become a profession. The question, however, is "what kind of jobs will there be?" nobody knows for sure, and the various economies are not even prepared because they do not know what to expect.

Chapter 3
Job Automation

A new study by PwC shows that 40% of jobs in the U.S might be replaced by robots in 15 years. The other developed economies do not face this risk since they have fewer jobs that can be replaced. In the United Kingdom, only an estimated 30% of their jobs face the threat of technical improvement in automation from robotics and Artificial intelligence. For Germany, it is 31% while the estimated percentage for Japan is 21%.

The reason why so many jobs in the US are threatened by automation is that a majority of the employees in the US work in positions where the tasks are routinized, like filling out paperwork. The industries that face the highest risk of the technological revolution are those related to retail, manufacturing, and transportation. How did PwC come up with these estimates? They broke down all the types of tasks of jobs in various industries. The researchers then used an algorithm that put into consideration the "automat ability" of all those tasks and the attributes of the employees hired to do them.

A good example of how jobs in the US are prone to replacement by automation than those in the UK, as per the research, is in the financial field. Despite the fact that these two countries both have identically service-dominated economies, jobs in the financial service sector in the US are a little bit more routinized and retail oriented. Financial services job in the UK, on the other hand, are occupied mainly by professionals who work in international banking. Their jobs are very difficult to automate, and they require higher educational levels.

A bigger portion of Germany's workforce is employed in the manufacturing sector compared to the UK's, which are exactly the kind of jobs that robots will be created to do in the future. This explains why it has a bigger percentage of jobs that faces the threat of replacement by automation. Compared to U.K, U.S, and Germany, Japan has the lowest percentage of jobs that face the risk of automation. Partly, this low percentage is due to the fact the jobs that are highly replaceable by robots are not so many in Japan. Take retail, for instance; it requires skills and far much more training in Japan—the workers have more organizing and management tasks compared to the same jobs in other economies that were studied. Also, Japan already has widespread automation. The researchers recommend several policy interventions that may be used to address job losses that might accompany automation. One of those policies is workforce retraining programs or the much discussed universal basic income scheme.

Chapter 4
Bill Gates and a Threat to Jobs

Output in the manufacturing sector keeps on improving, even though manufacturing jobs in the US are becoming fewer each day; and this trend will go on. Not only are jobs going overseas, but machines are now taking over jobs that humans once operated. This trend is only in its infancy; meaning that the momentum will keep on building in the years to come.

Machines are more efficient, less costly and do not grumble; for now. If you look at the situation from efficiency and cost cutting perspective, machines are better than humans. Consequently, this will keep on facilitating wage deflation. Sal Guatieri, from the Bank of Montreal, wrote a report titled "Wage Against the Machine," in which he said that automation is the cause for a weak wage growth. He wrote, "It's unlikely that insecurities from the Great Recession are still weighing, given high levels of consumer confidence." He went on to say that, "However, automation could be a longer-lasting influence on worker anxieties and wages. If so, wages could remain low for a while, restraining inflation and interest rates."

"The defining feature of a job at risk from automation is repetition." With this definition, many jobs are at risk—especially those that have been placed in the 'highly skilled category' such as X-Ray technician, radiologists, accountants, lawyers, etc." In 2016, North America ordered 35,000 robots which are 10% more than they ordered in 2015. China, on the other hand, made an order for 69,000 robots; Japan ordered 35,000 robots while South Korea made an order for 38,000 robots. For anyone that thought jobs are going overseas, here is the confirmation; machines are stealing them. Nothing will stand in the way of this trend; it is already in motion, and it is unstoppable.

The automotive sector has proved to be the biggest customer for robots. For robots ordered in North America, 20,000 of them were for the automotive industry. There was a time when humans did 80% of the jobs in this sector, today; robots have taken over that 80%. Within nine years, the number of robots in the US economy will increase by 300% as per the ABI research. It is not rocket science that more robots equal fewer jobs. A single industrial robot displaces six employees.

In 2010, a robot cost about $150,000 while today it costs $25,000—more than an 80% drop. The more prices fall, the more companies will acquire them. A time will come when robots cost less than $5000, and every business will be interested. Theoretically, machines increase the productivity of workers hence higher wages, but in reality, it is the opposite. Machines reduce the demand for human labor. The workers who are in big trouble are those that earn from $20 to $40 per hour.

Machines will take over from humans, and the cost of production will reduce by a great deal. Since unemployed people will have to compete for the available jobs, wages will drop. However, wages might rise in some particular sectors, especially those that demand a certain set of skills like robotics. Hence, the biggest issue in the future will not be inflation but deflation. Microsoft founder, Bill Gates, brought up an interesting argument during an interview with Quartz—the robots that will replace human employees should incur taxes that are equivalent to that of the employees' income taxes.

If a human worker that earns say, $50,000 is taxed, the robot that takes over the job should be taxed at the same level. According to Bill Gates, these taxes (that will be paid by the robot owners) would go a long way in funding labor force retraining. Former cashiers, drivers, and factory workers would be transitioned to education, health services, and other sectors where human input will still be vital. Gates argues that the policy will even intentionally reduce the rate of that adoption automation in some way allowing for more time to prepare for the broader transition.

The idea of what equals to a tax on efficiency may seem anathema to much standard economic wisdom. For a long time, the common line on robotization has been that displaced employees shift to more productive tasks, consequently growing the total economy. But that idea has already started to show cracks—as per Bill Gates, most people already believe that the arrival of that automation is a net loss, asking for a more active engagement with job retraining and other measures that target impacted groups. The effectiveness of job training institutions is still subject to debate.

Although Gates' thesis is in favor of the government's role in controlling the impact of automation, he gives two points that might be a little attractive to free marketers.

First, he says that the effect of artificial intelligence and robotics in the next two decades will be a much highly concentrated version of the stable incremental displacement present across the 20th century. The market on its own will not be able to handle the pace of that transition. He goes on to suggest that a big part of the potential for utilizing free labor in better ways will be in the public sector.

Secondly, and maybe more importantly, Bill Gates says that automation will not be given a chance to thrive if the public does not embrace it. According to him, it would be really bad if the people in general have more fear than enthusiasm about what this new technology is going to do. He goes on to say that, if anything, taxation is definitely a better way to the entire automation situation instead of just banning selected elements of it.

In other words, Bill Gates maintains that, if automation is not clearly an advantage to every member of society, then it might lead to some kind neo-Luddite movement and this will hinder technology much more severely than taxation would ever do. If you have a hard time believing

Bill Gates, just look around. The common belief that the benefits of globalization were unfairly or poorly managed has directly led to a political resurgence for fans of tariffs and walls. The same dynamic might end up repeating itself if the government and significant stakeholders do not roll it out wisely.

Chapter 5
Artificial Intelligence and Automation

When asked how he tells his kids to prepare for the future of working with artificial intelligence, Peter Norvig said, "I tell them... Wherever they will be working in 20 years probably doesn't exist now. No sense training for it today. Be flexible," he said, "and have the ability to learn new things.

Future of work experts and AI scientists believe that in the future there will be less full-time traditional jobs that require a single skill set, less routine administrative tasks, and less repetitive manual tasks—many jobs, then, will be all about "thinking" machines. From managers to janitors, everyone will adopt new ways of doing their jobs with machines in the next 20 years or so to come. One issue that is not clear, however, is whether the technological revolution will create more employment opportunities than it will destroy.

According to Al Toby Walsh, copying (Al computer) code costs almost zero and takes as much time. He goes on to say that whoever thinks technology will create more job opportunities than it will destroy is lying to themselves because nobody knows for sure. The jobs that AI will create will be different from the ones that will be destroyed, and they will require entirely different skills.

Hamilton Calder, CEO of Committee for Economic Development Australia, thinks that everyone should learn to code. However, Mr. Charlton disagrees strongly. He is confident that you need not compete with machines to be successful in the future economy. Professor Walsh argues that, even though machines will be far better coders than humans, for geeks, there is a great future in inventing the future.

It is time that people stopped encouraging the young generation to work towards a 'dream' job, says CEO of FYA, Jan Owen. Nobody should focus on an individual job. Instead, people should aim at developing a transferable skill set which includes; digital and financial literacy, project management, collaboration and the ability to carefully evaluate and analyze information.

ROBERT HILLARD, A MANAGING partner at Deloitte Consulting, believes that future work will be divided into three categories;

• People who will work for machines like online store pickers and drivers.

• People who will work with machines like surgeons who will be using the help of machines to diagnose.

• People who will work on machines like designers and programmers.

The human-machine teams will unite AI algorithms with human skills like emotional intelligence and judgment. According to Mr. Hillard, jobs will increase, but they probably will not be better. Those that will be working for the machines will have the most difficult time.

Yes, being human is a skill that you could leverage for income. Computers barely have emotional intelligence. The social jobs that need emotional intelligence (marketing jobs, being a nurse, being a psychologist) are safe. In the future, being human could be a job by giving services that machines cannot give—services in the caring economy, such as being empathetic. Some of these unpaid volunteering jobs could become "service jobs of love" in future.

Computers are not creative or imaginative. Surprisingly, some of the oldest jobs ever like being an artisan or a carpenter will be the most valuable ones. People would rather see something carved by a human as opposed to a machine. Even with all the preparedness for future work, Mr. Dawson thinks that everyone should plan for themselves. Develop the skills that will be needed and always pay attention.

Chapter 6
Auto Industry Jobs That Will Be Lost To Automation

Auto manufacturers have realized that automation will be cost-efficient than human labor in producing automobiles. Robotic technology means no paying wages and accompanying costs like sick leave, health insurance, and taxes. Billions of dollars will be saved, and this is a proposition that many companies cannot decline.

At the moment, robotic machines are already in place to manufacture parts and assemble vehicles. Some of the most recent technology in the manufacturing sector is computer-driven software which operates from blueprints that are pre-programmed. There is still a demand for machine operators, but the jobs are not the same. Below are some of the jobs that are likely to die when automation takes effect entirely. The Toyota Company gives a clear insight of how automation will take jobs from humans. Companies are adapting robots to produce crankshafts, axles, and auto chassis parts. These robotic assembly lines are reducing the demand for workers significantly. Quality of work done by robots is far much better than that done by humans (fewer defects and less waste). It will only take a short time for robotics to be perfected and fewer workers will be needed. This job involves putting parts together to create sub-components of vehicles on an assembly line in a manufacturing plant. Very soon, automated machines will take over these jobs. With minimal training, these entry level jobs pay around $21 per hour. Automation makes it possible for robots to perform most of these welding tasks. Brazing, soldering, and welding jobs are becoming less each day as technology advances in the industry. New job descriptions are going to emerge, and they will require additional knowledge of the machine being used in the automated process.

CNC MACHINERY IS NOW able to create any part, flawlessly and in bulk with the pieces being identical. With this kind of technology, metal fabricators will barely have work to do. Marketing jobs will no longer require humans, but they will rely on automated systems involving internet technology. People can now order vehicles online via automated ordering systems.

Newer automobiles have incorporated a computer system to generate error reports. When connected to a diagnostic machine, error codes can be read and hence diagnosis of the problem via a computerized system with sophisticated software. The demand for auto mechanics will never be at zero. However, most of the tasks they perform today will be automated. Just like a diagnostician, there is software capable of resetting the error code in a vehicle when a diagnostic has been run.

Automobiles are occasionally tested via automated systems to analyze their performance capacity and general quality. Computerized systems provide readouts of data for future analysis. Test drives are no longer necessary as analysis can be performed in the factory. Self-driving vehicles are coming. Human drivers will become a thing of the past. Taxi drivers and Uber drivers will be forgotten in a few decades. Delivery services will be replaced by self-driving trucks. The jobs that will probably be available are those of freight handlers. Automation is slowly taking over. Although it is still far from perfect, there is steady progress, and it will not be long. Very soon, about 2 billion jobs will be lost worldwide to automation when it finally takes over.

Chapter 7
The Rise of Automation and Coding

Automation has become quite a common term in the current IT community. It is everywhere; there is sales and marketing force automation, the automation of software in business processes, workload management automation among others. What makes it so important?

Automation means raising the standards of business traditional computing resources. It involves making manual and labor-intensive processes less dependent on detailed human intervention. Using aspects of technology advancements like sophisticated algorithms and machine learning, vendors can provide companies with the ability to automate all types of key processes and tasks in many useful and exciting ways.

Almost every system administrator has a large number of benefits to reap from automation. Automating activities creates time for people to focus more on other key aspects of running a business.

For instance, in a virtual network environment, a person who spends most of their time checking every detail of a virtual machine performance, will be able to pay more attention to the big picture and how they can scale the system more efficiently. The automation software will be responsible for the other small details.

The benefits of automation can be very diverse. Take for example of someone who runs mechanical hardware or software processes in a manufacturing environment. Automation can take care of processes leading to quality assurance, production tagging or any other area and the human manager will be free to do other important things.

Some automation tools can perform managing tasks in an enterprise environment. There is a tool called Zapier that helps in integrating apps easily and also monitors the automated information that flows between them. It provides a way to repeat pre-built processes without having to reinvent the wheel.

These types of tools have a profound scope of applications. If this kind of automation is connected to Android, Apple, and other platforms, human managers will have the advantage of adaptive automation assistance.

Stackify, another useful tool, has several benefits. It can help find hidden exceptions, synchronize business activities or solve application performance issues. For instance, a ticketing system or a booking/ reservation system uses several different applications. Companies strive to assess how data flows between those applications. They would appreciate product support and an engaged model for fixing glitches and bugs.

In such cases, Stackify can make problem-solving more proactive. It will not only provide notifications when something is not right but also perform an overall analysis of those applications as they develop, to catch any issues. One principle, DevOps, combines development and operations; making the development process more agile and fluid. It helps in automating some of the processes that bring applications into full swing. These are the types of processes supported by Stackify.

Companies that are barely conversant with these functionalities should be focusing on how to incorporate automation into their business practices. Be it a service or product business, and whether the process concerns customer connections or manufacturing, various kinds of automation may help improve the outcome of businesses.

Chapter 8
Cyber Security

Cyber threats continue to increase in number and strength. The future of cybersecurity looks more challenging and complex. Organizations are relying on automation and analytics to help cyber specialists do their job.

Although the field of cyber security can be complicated, some of its aspects are very clear. The number of threats facing large organizations is rising frantically, just as is the number of bad actors creating them and also the number of systems at risk from these cyber-attacks. According to Statista, there were 22.9 billion connected devices in 2016, and this number is expected to hit 50 billion by 2020. The Internet of Things (IoT) will come up with numerous needs and problems for cyber security as more devices come online. There are increasing cases of data breaches. How can organizations cope with this scary growth trajectory?

For some businesses that are subject to huge numbers of entities, the ultimate approach is to apply automation and analytics. For instance, in customer analytics, segmenting customers by their value, focusing on the more important ones and knowing what they are likely to buy is the typical approach. They can customize automated offers to suit each customer's preferences. These same technologies can be used to save cyber security from its growing predicaments. The cyber specialists in organizations today are not enough to combat the number of threats and this imbalance is only expected to get worse.

Cyber security is usually reactive to breaches and hacks, with action being taken after the problem has occurred. The technology common in addressing cyber-attacks uses "threat signatures" based on patterns of the previous attacks. These approaches are of very little help in preventing new kinds of attacks. An efficient solution is to use analytics to screen and predict threats and take automated corrective actions. Because cyber security issues are sensitive, humans will still be needed to investigate and confirm threats, especially internal ones. However, their tasks will be much easier and productive.

Other real-time and predictive approaches are emerging from software vendors. The same modeling and software approach employed to detect credit card fraud—a kind of anomaly detection—is being used to monitor behaviors in cyber security attacks. They can detect emerging anomalies faster than threat signatures, and they may prevent a significant breach before it occurs.

Technology might not be able to deal with every cyber security problem. Although organizations may undertake some automated actions, they will want to investigate issues detected by analytics before they take corrective actions. These investigations involve testing, research and in some cases, interviews for internal threats—and all these require human experts. Therefore, the most efficient cyber security environments will be highly complex hybrids of machine and human intelligence.

It will also involve an adequately defined process of detecting, screening and combating threats that outline the roles for capable humans and smart machines. This process should just focus on identifying and qualifying threats but also on taking rapid actions. This will not be easy, especially with a large number of threats; however, analytics-based prioritization could be of help.

Chapter 9

Consumer Automation

For someone with a Wi-Fi thermostat, all they need is a computer, tablet or smartphone to control the temperature in their home. Energy efficiency in a home is essential, just as in the factory. The Internet of Things (IoT) is coming up with smarter ways of controlling appliances, HVAC systems and tracking the safety of an entire home. Homeowners can even be in control while out for dinner, in the library, and at a ball game.

The increasing use of smart home apps and devices shows the growing popularity of automation in making homes safer and more user-friendly. Consumer robotics such as pool cleaners, lawn mowers, and robotic vacuums has recorded a high number of purchases. A market research by Tractica shows that the number of sold consumer robots will rise to 31.2 million units in 2020 from 6.6 million in 2015. This is because consumer robots cater to several different needs and smartphones being popular, you can control them remotely in a snap. Pressing an app is no more difficult than dialing a number.

Controlling a thermostat is just a part of a broader strategy of home automation. Window sensors, wireless doors, lights and other appliances can all be programmed for remote control. Energy-conscious consumers who do not like inaccurate ovens and glaring inside lights can set their space into a mode that is perfect for them depending on whether they are away or home and the time of year. Since there is no way to make dirt disappear; scrubbing, and vacuuming floors will always be there. Robotic vacuum cleaners are designed to remember a home's layout for efficiency and then go back to the charging station on their own. This is far much better than industrial robots which were not able to perceive their surroundings.

According to an article by the Robotic Industries Association, "Intelligent Robots: A Feast for the Senses," their behaviors are "uniform and predictable. No variations, no deviations." Software and sensors are offering higher intelligence to both household and industrial robots. What do you think about a robot personal assistant with artificial intelligence? It can tell bedtime stories to your children; recite recipes, monitor the house when you are not around and dim the lights. A certain article by Techcrunch.com, "Robotbase Wants To Put an Intelligent Robot in Every House," gives an insight on the personality and capability of this robot. Advanced algorithms are incorporated into a home's connected devices to allow the assistant to take in surrounding data. The same technology being used in the workplace is the same one being used at home, but for different purposes. When one sphere develops, the other sector benefits, they both have a lot to reap from intelligent technology. The main purpose of a robot personal assistant is to handle tasks discreetly in the background and proactively alert you using sufficient artificial intelligence, advanced computer vision and plenty of sensors.

According to experts, it is clear that robots will be a part of life in 2025. These machines will be in our homes, stores, places of work, you name it. The only question that everyone is trying to answer is what this new change will leave in its wake as far as humanity is concerned, especially when it comes to the workplace. More than half the percentage is convinced that technology will not take over more human jobs than the ones tech will help create. That is fifty-two percent of people optimistic about machine intelligence transforming us for the better, with the other 48% fraction still skeptical that this will spell doom for jobs and wages. The issue is still subject to confirmation and lengthy debate. But there is a catch since there is a possibility that most employers will still consider human labor for white and blue color jobs that machines can do well.

According to the people who believe we are still safe from machine takeover, technology will still play the part it has always been part of from before. Machines will thus help in improvement of productivity as well as creating the opportunity to focus on human jobs that require skills for problem-solving and advanced creativity. For others, however, this will remain a precariously economic rule whereby automation poses a significant risk to the increase of corporate profits and wages. Machines can carry out more intricate and complex tasks nowadays. This is all thanks to the advancing computer vision, sensors, ability to learn, and algorithm among other technical advances. According to an Oxford study, 45 percent of jobs in the U.S is most likely be automated in the period of two decades. However, experts have varying opinions concerning the threat that machine intelligence poses to humanity regarding work. One of the things that optimistic experts point out is the fact that automation has never posed any threat to reduce job opportunities in the economy so far, and the same is highly unlikely in the future. This is because automation helps reduce prices, hence increasing demand for services and goods, eventually creating jobs.

The other factor that is termed reliable is the ability of automation to create more jobs than it can displace. This difference means there will always be enough jobs created in the process to ensure no lack of job results. Others say that the main risk to jobs is not necessarily AI, but large-scale shifts in employment to areas where labor is less expensive. And despite the fact that automation has been feared to take over human jobs for the past few decades, the real risk lies in the aspect of risk management rather than technology taking over human jobs. The experts who do not believe that this will not mean any threat to jobs have concerns over the versatility and always advancing capabilities of AI and robotics. They claim that this will not only affect numerous economic sectors but might extend to have effects on whole swaths. This comes down to economic efficiency, and the trend is already affecting some sectors. Others believe that automation processes that target the core of the economy will kick humans out as the shift takes effect. For some experts, only the most educated humans will stand a chance in the war for jobs against machines. These experts believe that the worst part is that students are not being well prepared for this change. As much as automation may seem like a threat to the survival of humanity as far as employment is concerned, it is important to remember that this technology can come with jobs never before created. It is more about looking into the jobs that will be replaced in the future because bots were not able to create it and will be lost to machines.

Chapter 10

Automation in the Healthcare Industry

Automation is growing in popularity every day in the technology world. And this is not without reason. Almost all forms of technology are gaining a lot from automation. It is present in just about every area; from social media to digital marketing to software and IT. The healthcare sector is no different; it is quickly embracing automation. It is impossible for the healthcare sector to remain static. It has to keep changing to enhance the livelihood of the human race; which is one of the primary reasons for embracing automation. Automation has so much to offer to the healthcare industry.

Most hospitals need a lot of staff because there is no way to automate some of the processes. Once the process of automation finds its way into hospitals, labor costs will decline to a large extent. A reduced labor cost does not mean that there will be a massive layoff by the hospitals; rather, the hospitals will only need the staff for extra services—for example training. Therefore, the hospital will be able to offer additional services with the remaining funds. To run a hospital, you must know the details of all healthcare activities. So you are undoubtedly aware that there are activities that have to be repeated every day in the healthcare field. Nobody likes to keep doing the same thing repeatedly. Enter automation. You only need to automate the process and have them repeated daily with no manual intervention.

One of the essential reasons for automation is increased perfection. With human staff, you will always find several errors. Regardless of how efficient an employee might be, trust them to make a few mistakes. In the long run, these errors might cost the hospital a lot—and in the worst case, a life. When you automate a certain task, you reduce the probability of such errors happening and make the healthcare procedure more efficient and effective. The entire healthcare process involves numerous parts, and one of the most crucial parts is the diagnosis of a disease. A lot of reports are required to perform this action, and it can be implemented best via automation. Automated reports are usually delivered in pre-set formats, specially created to ensure that a physician makes the most of them. This leads to a better analysis of reports and enhanced diagnosis of diseases.

Healthcare services should be of the highest order. The healthcare provided is usually depended on data acquired about a patient's condition. Therefore, if better data is obtained, health care services will improve. You can only acquire the best information with real-time data. Automation can show a lot of important data about a patient in real-time. Automation has become an important part of technology today. Many industries are embracing it every day. Automation in healthcare has improved diagnosis and treatment of diseases. Automation is and will continue to affect the healthcare industry significantly.

Automation is quickly killing many jobs in the healthcare industry. That is good news for insurance companies—the middlemen with the upper hand over patients and doctors in the third party payer system. IPads and computer terminals are already taking up the jobs of front office staff, where patients check in and out. Next, the robots will take over the back office where patients are directed to various offices. The doctor's office will only be left with a nurse or doctor to read the chart and refer or discharge patients. The paradox is that patients are not the ones that stand to gain from this office automation. Health insurance companies stand to gain the most since they will deduct the payments to doctors to mirror the savings from doing away with the front and back office staff. Health insurance companies, by reducing the fee they pay for patients visits, will give the doctors no alternative to running their office the new automated way. This is the reason why automation is a cause for celebration for insurance company stakeholders. Fewer payouts to the doctors will raise bottom lines and equity prices.

Chapter 11
AI Is the Future of Cybersecurity

Very soon, as Artificial intelligence systems become more advanced, you will start seeing more sophisticated and automated social engineering attacks. With AI-enabled cyberattacks, there will be an explosion of network penetrations, a wild level spread of intelligent computer viruses and personal data thefts. Ironically, the best defense against AI-enabled hacking is AI itself. However, the chances of this leading to an AI arms race are very high, which may result in very troubling consequences in the long run, more so, when big government actors take part in the cyber wars.

This research focuses on the intersection of cybersecurity and AI. Particularly, how AI systems can be protected from bad actors and how people can be protected from malevolent or failed AI. This is all a part of a more significant framework of AI safety. What can be done to create exceedingly capable AI which is also beneficial and safe? There are so many articles trying to portray what problems might come with "true AI," either as a result of a programmer's error or as a direct impact of the inventions. However, nobody seems to be addressing AI hacking and intentional malice in design. On matters pertaining purposefully unethical intelligence, anything can happen.

Bostrom's orthogonality thesis suggests that an AI system is capable of having any combination of goals and intelligence. The goals can either be introduced through hacking or the initial design or later; as in the case of off-the-shelf software—"just add your own goals." Therefore, depending on the organization for which the system is bidding (terrorists, corporations, military industrial complexes, dictators, governments, sociopaths, etc.) it may try to cause damage that has never been seen before or one inspired by previous occurrences.

Even today, it is possible to use AI to defend and attack cyber infrastructure, and also increase the attack surface for hackers, that is, numerous ways for hackers to enter a system. As the capability of AI increases, they will first arrive then overtake humans in all areas of performance, as has already happened with games like go and chess, as well as important human tasks like driving and investing. Business leaders should understand how complex the future situation will be from the current one and how to plan for it. If a cybersecurity system fails today, the damage will not be good, but it will be tolerable—someone may lose money or privacy. But for advanced AI (human level and above), the damage might be catastrophic. Some tech giants like Bill Gates, Stephen Hawking and Elon Musk have already expressed their concerns about the possibility of AI evolving to a point where humans will not be able to control it.

Today's cybersecurity system gives you a chance to perform better if the system fails, unlike with a SAI safety system; you either have a safe, controlled SAI or not. Cybersecurity aims to minimize successful attacks while focusing on eliminating successful attacks. Business leaders should acquire knowledge on the cutting edge of AI safety and research on security (even though, for now, it is nothing different from the state of cybersecurity in the '90s), and the current situation with the absence of security for the internet of things.

Chapter 12
The Future of Automation

Many firms are turning to automated workers; from hospitals to stock markets, to law firms. So how is this process going to affect the human workforce—or you, for that matter? You have heard so much about how robots will take your job away from you, but what is going to happen? Whose job is at risk? How will your workplace look like in about five to ten years? This is what experts have to say.

According to reports, 47% of US employees are at a risk of having their jobs automated. The same case applies to 35% of employees in the UK. The threat is even higher in developing countries as two-thirds of workers are at a risk of being replaced. The issue of automation stealing jobs is not exactly new. A finance professor at the University of California, Bhagwan Chowdhry, says that this is not the first time that automation is happening. He talks of the shifts that occurred during the industrial revolution. Automatic looms and other machines replaced human weavers in factories.

The main difference this time, however, is the fact that it will affect not only blue collar workers but also many other white collar workers. The common thought is that low-skill, low-wage jobs like cashiers or warehouse workers are at the most risk. Nevertheless, robotization may affect middle-income jobs like inspectors, junior lawyers, security guards, office workers, chefs, and clerks. According to Carl Benedikt Frey, the transition pains are of more concern. The jobs that will be automated will need a different set of skills from the ones being created. The primary challenge is ensuring that the displaced employees find something else to do. Is it the responsibility of companies pursuing automation to help replaced employees acquire new skills? This is beyond the company's alone—it should start in school. The current education structure maybe unfit for purpose in an environment where technology is rapidly changing. One primary problem is that the training, education and political training institutions are not being updated to keep up with technological growth; and a lot of people will end up being left behind. In the workplace, employees will also continue to attain new skill sets instead of using the old ones throughout their entire career—which could become obsolete. The difference between learning and work might become more amorphous. With the current dichotomy, those who work will not need to learn and those who learn are not working. Now is the time to reevaluate the traditional five-day work week and adopt one where they work 60% of the time and learn 40% of the time.

According to a particular research by McKinsey and Company, less than 5% of jobs can be fully automated by the current technology. This is because; most occupations are too changeable and varied for robots to assume all the tasks. 60% of jobs could have a third of their activities automated. So technically, many people will keep their jobs, but they will have to do them differently. Robots will not replace you; they will compliment you, learn how to work with robots. There are many tasks that machines cannot perform at this point in time. There will be a rising demand for negotiating, caring, and nurturing skills among others. Machines could handle the tedious and repetitive habits and make humans' jobs more human.

Today, a good number of parents would not like their kids to follow in their career footsteps; not because they are unhappy or unsuccessful, but because they feel that future job prospects in their areas of practice are bleak because of automation. Automation includes both artificial intelligent software programs and mechanized robots. It is expected to eliminate 6% of US jobs in the next five years. Everyone needs to be worried; not just low-skilled workers.

The insurance industry is already feeling the effects of automation. Fukoku Mutual Life Insurance in Japan has, in the recent past, replaced 30 of its medical insurance claims representatives with an Artificial Intelligence-based system based on IBM's Watson Explore. The software can examine and decipher data, and that includes unstructured text, audio, images, and video. First it was the ATM that took human banking jobs, followed by the smartphone app. Most probably, the remaining human-based representative and teller banking jobs will be rendered obsolete by Al, not only will Al conduct cash transactions, but it will also process loans and open accounts for less than half the time and cost.

Financial analysts were once thought to be indispensable to a company, given their keen eye that could spot a trend even before it happened, enabling institutions to make adjustments to their portfolios and make billions of dollars. However, they cannot even come close to Al financial analysis software efficiency. Financial analysts' jobs may even be the worst hit. Manual labor jobs will not be spared by automation. Robotic bricklayers are almost being introduced to construction sites, and each one might replace two or three workers. SAM (Semi-Automated Mason) can lay up to 1200 bricks in a single day, compared to a human worker who can only do 300-500. Human workers will still be needed to work with SAM, but the machine will lead to the laying off workers.

Very soon, a robot will replace the employee restocking goods in the supermarket aisle. And since the robots are quite advanced, they can take over tasks that previously needed a pair of eyes like managing inventory on the store shelf. Artificially intelligent robots that carry out almost every task on a farm are very likely to replace farmers. In Germany, a family-owned dairy farm becomes among the first to incorporate Voluntary Milking System robots. Cows walk up to the machine when they want to be milked. Traditional taxi drivers are already feeling the pinch from services like Lyft and Uber. However, the three are about to watch their jobs disappear as autonomous vehicles take over. Many politicians love to say how they restore manufacturing jobs in the US, but that is not about to happen. Even China will see robots take over jobs that once belonged to humans—as a matter of fact, that is already happening. It turns out that AI can do a pretty good writing job and future content sites might do just fine without the help of human writers. It is also possible that you start seeing actors looking very young in movies; just as they did when they first appeared on the screen. To some extent, it might become possible to resurrect deceased actors from the dead with artificial intelligence.

Computers have developed over time to reach today's advanced technology that you can count on to carry out a large number of functions. This trend is called the digital revolution, but the reality is that it is becoming a part of our everyday life. From our basic theories to language and perception of how things operate, computers are having a major effect on human life now than ever before. Most people believe that this means the world is more of a computer, which can be coded as computers are. But computers are taking things to another level with an advance in the aspect of artificial intelligence. One of the factors where computers are elevating is coding. Coding is perceived to be logical, hackable, and the destiny of our advancement. This is because coding is more than just about computers, but more of the design of life form itself. You might like this reasoning or think it is not reliable, or whether you are among the coding elite who are taking the lead from phones to computers, you don't have to worry. The good news is that computers are becoming more intelligent nowadays than they ever have. It is believed that these machines are starting to take a completely different turn with a new language, which is proving difficult for the best coders to crack. Recently, some of the biggest companies have taken to the forefront in the pursuit of machine learning. Formerly, engineers would use conventional programming where one writes explicit sets of instructions that the computer can follow. But machine learning is different and better. This approach does not include encoding instructions; it is more about training the machine. For instance, in case you are training a neural network on recognizing anything, all you need to do is present photos of what you want it to recognize. If this does not bear results, you don't have to record; all you have to do is present it with more pictures. And although this trick has been around for decades, it became more powerful only recently. This has resulted from the emergence of significantly distributed computer systems and neural networks. These machines are already taking over and controlling most of our online activities.

Some of these machines can learn to convert some languages instantly. Other forms include autonomous cars, which make use of this technology to prevent any accidents. Another type that uses this type of machine learning is the Google search engine. This new way of life has outperformed the old human-written rules. This means the role of conventional programmers has been overturned. Rather than taking control of creating the laws and feeding them into these machines, programmers now can only train the machine like pet trainers or parents. And just like any parent would direct their young, or a dog trainer would instruct their dog, this is all about what you need to do with these machines. Machines are portraying every aspect of human workers in the workplace. And with these advancements in place, it is clear that automation is here to stay. In some companies, robots are taking over already, and the change seems to blend in quite well, which only indicates just how easy machines can take over jobs without inconveniences. These changes only mean one thing; we need to prepare for the future of automation because it is already here with us. But how do we get there? For many companies, this has become an irresistible strategy, and these firms are making the most out of tech with high-end products. But despite the high number of robots already in the workplace, the future is about man vs. machine. The best thing to do is look for ways in which humans can work with machines, which will help in making organizations intelligent and offer improved performance.

It is clear that most upcoming business models are finding it profitable to take advantage of the benefits that automation offers. Some of these establishments are making the most of machines to deliver value without necessarily relying on human based services. Most of these companies are leveraging on technology to create networks that form the foundation of the organization's success. This helps create revenue, not to mention the efficient use of assets. This also proves that going robotic can always come with unprecedented advantages for startups. These aspects are rapidly becoming the new workforce models, most likely the essential approaches to understanding and adapting to automation. These models are coming in handy when it comes to getting the best out of one's workforce, and most organizations are adapting to the new way of things. But this generation of talent requires an input of diverse work environment and work location.

Since automation is increasingly gaining ground in the workforce, the best thing that humanity can do is preparing for the inevitable, uncertain future. These changes are already altering companies' needs for space as well as the location of facilities and configuration of space among other aspects. Similarly, companies need to adapt to the need for increased flexibility, technology-readable facilities and access to new talents, to mention but a few. Apparently, automation is here to stay, and it only gets bigger, better, and stronger. Businesses have to remain agile enough and consider other aspects as well to retain a competitive edge in the market.

Chapter 13
Colleges: Jobs of the Future

This has always been a question in the minds of employers and educators: will automation entirely replace human occupations in the future? Elon University partnered with Pew Research Center to find out whether technology analysts, business leaders, and educators think this will happen and what higher education should do to ensure that humans stay in the workforce.

To compile the report, Elon University and Pew Research Center surveyed 1,408 tech experts. The experts consisted of a wide variety of digital ethicists, internet researchers, academics, policy professionals, and members of civil society organizations. Among the respondents, two-thirds said that education and job training might be successful in training employees for the increasingly automated economy. A third of them said that they are not confident that education and job training will change quickly enough to suit the labor market demands. A number of these skeptical respondents even said that they were worried about generating the required political will to begin a large-scale reform of the economy and education.

Based on an article from the New York Times, for instance, a principal researcher at Microsoft Research, Danah Boyd (she is also the founder of Data and Society), says that she is utterly confident in the ability to identify job gaps and create educational tools that can address these gaps. However, she says that she does not trust the US to have the political will to deal with the socio economic issues that are underpinning skills training.

More pessimistic respondents have their attention on the limits of human learning. Andrew Walls said that "barring a neuroscience advance that enables us to embed knowledge and skills directly into brain tissue...there will be no quantum leap in our ability to 'up-skill' people." According to the two groups of respondents, employers and higher education alike, do a better job educating students in both soft skills and technical skills.

Artificial intelligence has seen contradicting reception on whether it is helpful or destructive. For some, this technology is a destroyer of jobs, while others take it as a way to liberate workers from administrative tasks that do not comprise of much creativity. With the capabilities of AI increasing by the day, we can see it as a means of relieving our need for thinking too much, or a tool that helps our brains to perform more creative thinking. If we are to overcome automation, the best way is looking at it as a tool that can help us think more creatively, lest the technology overtakes us and takes the lead over human workers. And since coaching cloud is here, there is still every chance for humans to win this race.

A coaching cloud employs a strategy whereby machine learning helps workers in doing the job for effective results. This method is based on software for gathering data from a diverse network of workers then identifying the ideal technique for doing things. Perhaps the best part is that this software will come in as a real-time coach that can guide employees to attain a better outcome while collecting data that can be fed into the system.

Even better, this approach is not about one design for all, but rather each coaching is tailored to fit each worker and matches the task one is doing at any particular moment. The software advances with time as it learns the most appropriate practices declared effective across the board in different situations. Besides, it also makes the most of the isolated cases where creative workers have come up with new and more practical solutions, and then the software can add those as well into the coaching process. With this strategy, other workers can learn from the most creative ones in the group, which helps humans attain mutation to engines within the evolving process. Thus, ideas are gathered and used to benefit everyone.

What's more, this method can help in solving problems as well, especially the easy training of remote workers. In the end, someone's work becomes more significant than their efficiency in the job. It is also far-reaching in reducing the complications with how fast someone is to adapting to new skills or improving the existing ones. This strategy can be embraced in different economic sectors like sales and service jobs that require human-to-human interactions and more of interpersonal communication.

Some of the crucial things to remember here include going deep rather than big with data. The step involves zeroing in on a specific domain and pursuing that instead of pacifying on a wide range of fields. Besides, paying more attention to the data and not much on the algorithms can help. This means using the normal open sources than sinking to designing unique algorithms. Above all, the data should be made useful and, more importantly, visible too. This necessity calls for the creation of user experience that can encourage the use of the data and make sure it is easily accessible to users.

Chapter 14

Automation and Perception

Researchers are hot on the pursuit of understanding and predicting the effects of machine intelligence on work and jobs as businesses delve into the use of the technology. In essence, the concept here is all about machines learning through experience and improving performance with time. According to optimists, once machines have taken over cognitive, labor-intensive chores, humans will be able to take on the more creative tasks as they will be free. It is also believed that humans and robots working side by side will be far-reaching in triggering imagination for better achievements.

According to conventional belief, increase in job performance goes hand in hand with increased value to the company, but the theory has been debunked. The new perception is that the line is drawn depending on the variance in role means all the difference. Nevertheless, despite the fact that AI, digitization, and technology are influencing job changes, the relationship between value and performance still holds. This aspect only becomes more complex and with more opportunities for creating value. In this case, the return on investment, also known as return on improved performance, is used to measure the value offered by enhanced performance in particular position. Most at times, humans are critical of their fellow humans making errors in the job, but the idea of a machine getting something wrong and ending with mistaking freaks them out all the same. For instance, if an autopilot or autonomous car is involved in an accident, people will be afraid of the technology, but humans have always caused accidents for sheer negligence. It all comes down to trusting humans in most cases, while many of the times, machines are usually more precise and accurate than humans can achieve.

This means putting trust in robots learning and being able to execute duties might take a while before humans appreciate the machines. So companies should focus on several steps to rethink the issue of artificial intelligence and automation.

• One of the things to be keen on is the difference between proficiency and pivotal roles in the organization clear.

• Then understand the underlying factors between the performance and value for the proficiency and central aspects.

• Besides, you need to disaggregate the various aspects that determine how artificial intelligence can be useful.

• Then determine the particular activities that might be transformed by the different AI forms as well as keeping an eye on the capability, the related cost and the risks.

• After that, plan on how you can engage stakeholders in the process of understanding as well as embracing the possibility of applying changes to the work reorganization of these biases and don't forget to factor in the resistance aspects as well.

To get a reliable, competitive edge, leaders in organizations have to understand the extent to which AI and technology can affect the performance as well as the value equation. To be a successful leader, someone in the field will need to learn how to translate the pivot points in the business models to achieve specific implications for multiple projects. Besides, looking beyond the jobs and having an understanding of the transformative role can be far reaching too.

Chapter 15
Manage Automation and Jobs

I t is clear that automation has seen widespread recognition so far. It is about time executives found a way to think of the means to decipher just how the technology can have specific effects on their organizations. The best approach, in this case, is not considering the type of jobs in the risk of being replaced, but looking into what can be changed and how. Forget the job or the technology aspect, considering the work itself goes a long way. As in the regular jobs aspect in organizations, work will always be there. The only thing that makes these jobs significantly fluid is automation, while the amount of work continues to increase beyond the traditional confines of a job. Employers should thus deconstruct the work into discrete elements if integration of humans and automation is to be achieved. This means seeing any task in a job independently and consisting of fungible components. This approach reveals a more efficient, impactful and effective combination of human and automation.

The fact that most AI and robotic applications take over the white and blue-collar jobs means the non-routine tasks are the only positions left to humans. This compromises organization's ability to keep human work, hence, reconfiguring the non-routine activities can help achieve new and different kinds of jobs. This comprises of robotic process automation (RPA), social robotics and cognitive automation. RPA is about automating low complexity, high volume, and routine administrative tasks, usually the white-collar type. The use of this approach can help increase accuracy and reduce costs as well as outsourcing on several regulatory processes, but this is only applicable once the work has been deconstructed.

In the future, global work systems will be able to provide alternatives to work arrangements such as the three automation solutions. This further comes with human work sources like contingent labor, talent platforms as well as traditional employment among others. Since the work created for humans in this method cannot be sourced solely through the job, getting more opportunities from within the organization goes a long way. Such positions can be created as deconstructed units, which can be engaged through numerous sources. Working with a combination of reconfiguration, work deconstruction as well as automation comes in handy when it comes to redefining the concept of leadership and organization. You must consider the body as a hub as well as a capital for the work providers' ecosystem. These providers comprise of Artificial Intelligence and automation, not forgetting the human sources like contractors, employees, volunteers, freelancers as well as partners among others. However, to bring out this combination at its best, you have to focus on the issue from the perfect approach as aforementioned. Artificial intelligence should come in handy in empowering global workforce, not to mention that it will have a considerable level of effect on the same. This may not be a regular occurrence in every job, but it will happen quite often in some cases. Leaders thus need to embrace automation strategy to realize the advantages that it comes with while keeping unnecessary costs at bay and getting a better work understanding.

Artificial intelligence is already having a perceived improvement in almost every day's aspect. And one of the factors that this change is driving over the top is the commercial part of society. Recently, Uber saw its first technological achievement when it unveiled an autonomous truck, which is an indication of the increasing shift in the economy. This could come with a significant impact on the political and economic arena. Although it is still not clear how long it will take autonomous trucks to take over, it is evident that this trend is coming on board, and probably with significant economic impact.

One of the most affected areas is the employment sector. This tech is posing a risk to loss of millions of jobs, according to a report by the former U.S administration. On their side, economists claim that automation is to blame, more than globalization would be blamed for the decline in the manufacturing sector. According to this report, the invention of artificial intelligence started back in 2010, which is when machine learning started picking pace as well as an increase in the use of big data. This also came with a high level of computation power. The report further states that use of AI and automation could help in enhancing growth in the economy through the creation of new jobs besides boosting efficiency in business. This trend comes with a fair share of downsides as well, which includes the destruction of jobs as well income inequality from similar issues. Automation is blamed on displacing several high-skilled jobs.

It is still unclear how AI and automation could affect job opportunities in the future. There have been predictions about the jobs that might be changed, and to what extent. Inequality is among the most affected areas, where AI and automation are feared to replace human jobs. Policies by the government might however not be able to offset the effect of such a trend. Such improvements like better training of the workers affected by automation taking over their jobs as well as increasing financial protection for those left seeking jobs are among the most recommended solutions.

Although AI is coming with a fair deal of craze in the market, the tech is still in its infancy stages. Autonomous vehicles might be ideal for a sunny day but might not be reliable in an uncertain environment. Besides, the tech might not be as efficient in cases of emergency. Not to mention, although autonomous systems can handle complex patterns in vast amounts of data, most types still lack common sense or the language that would be used by young kids.There are still significant technical challenges with automation. The government and private industry must address issues related to employment and paying attention to the technical aspect to make the most of AI and automation regarding the economy. As much as this tech is significantly taking hold, it still has a long way to go, but when it does, much will be at stake.

The healthcare industry has seen significant advancements in recent years, and still an extensive level of training data yet uncharted. From tech involved with taking patients' lungs, CT scans from different angles to automating dispensing expert radiology. There is still much benefit from automation with the ability to get more efficient and accurate health care. Since deep learning and other types of AI are already taking root, machines are replacing most white-collar jobs.

Machines are carrying out most types of work now than any other time in humankind history. Besides, robots have more intelligence now than ever before, taking over the work done by information workers among other duties. The aspect of a job being able to go to the automated end is more a question of whether it is consistent rather than an issue of being white collar or manual. This is all thanks to machines having advanced beyond performing manual tasks to take on cognitive functions.

This is being seen as a menace to the high-end workers like trained radiologists, who face the risk of being replaced by machines in their professions more than having their assistants take over. The question that remains eminent is on which jobs are in the line of the advancing automated technology. According to a study conducted several years back, almost half of the American jobs are at risk of potential automation, mostly the logistics, transport, and office support duty sectors. Besides, workers in the sales and service field also face a considerable risk of being replaced by machines.

So far, economists are worried that job polarization is taking hold. Such cases include middle-skill jobs like in the manufacturing sector declining while those in the high-skill category are advancing. Eventually, there is a division of workforce into two groups. The skilled workers like senior managers and architects are highly paid while such unskilled workers as cleaners on the other end are underpaid.

Previously, workers could shift from routine jobs in a particular industry and join the same kind of employment in another sector. But with the new wave of automation, big data is offering companies all they need to improve marketing and customer-service as well as the raw material to enhance machine-learning systems. This trend is leading to machines taking over the jobs that would otherwise be performed by people. Technology can help in many ways as software could outperform humans. For instance, software could sort documents faster than a clerk can achieve or writing editorial content more quickly than a journalist is capable of doing. While in the past technology used to create more jobs than it took away, the reality is very different now. This will result from the fact that automatic in a particular task could mean it will be done faster and more efficiently while still cutting costs. This means human workers will only need to do the other jobs available that are yet to be automated. Although the field of AI itself will come with a considerable level of job creation, it is undeniable that it will end up taking more space than it would create. It is clear that this will lead to decreased job opportunities in many sectors.

Chapter 16
Automation and the Future Economy

Technology has come a long way, and it has been here for well long enough now, with inventions moving from one leap to the other. The menace of having to deal with a future where jobs will be a problem has been around for almost as long as the technology has. When you look into our economic past, it is apparent that the trend of technologies taking over human work is bound to continue into the future. This might happen soon enough, and with significant disruption to the economy. But could there be a way of looking at the entire issue from a different perspective? Say in a case where future jobs are automatable rather than automation taking over? Even though AI means robots are growing clever, which the machines have done over time, it is, still obviously some tasks are still beyond the cognition of robots. And some of the most straightforward tasks have proved difficult enough for robots so far.

Quite apparently, since you cannot expect these machines to perform as humans would, it will be safe to say humans might find it difficult to understand just how difficult it may be to carry out these tasks. But this is one of the most crucial steps to start with since machines can be a reflection of the current economic system. On the flipside, manual jobs remain the most poorly paid, and these include duties like bussing tables, emptying bedpans, cleaning hotel rooms, and other similar tasks. But robots are best at the middle-class level jobs like accounting and payroll. The level of technology contribution in the widening of the income gap in the U.S. remains a controversial issue yet. For some economists, this is just but another factor, but for others, this is the bottom line. When technology and unemployment are on a collision course, there could be a severe problem for everyone, and policies should come into place. However, if we take it as unrealistic to think that intelligent machines could be stopped, then it is more likely as impractical to believe that smart policies could save the day. And the sad part is that once it is all lost in making a choice for the right approach to use robots, recovering could prove to be a challenge.

It is, thus apparent that as much as technology has seen humankind through a maze of many achievements, if not well handled, this very aspect could spell disaster for the human race. Embracing an approach to bridging AI and workforce to get the best results could be the best solution. This would help more than trying to fight the intelligence of these machines in hopes of making things any better down the road. However, it all boils down to the approach that society goes for, which could either make or break the effect that AI and automation could have in the job environment. Since automation is threatening to take over what workers have been clinging onto all this while, there is the need to find a way to go around the inevitable problem. The current state of events is putting future generation chances for jobs at a considerable risk, and it is apparent people are becoming increasingly worried over the advance in automation. This is imminent since there is need to replace jobs that people have lost along the way. Lack of this measure could lead to an identity crisis and other adverse effects.

The focus to this menace is about asking what humans can do with the help of better thinking machines than asking of what human jobs that robots could take over. It is thus essential that the risk of automation is reframed into an opportunity for augmentation, but what is the latter includes. The concept behind development is all about looking at what humans are doing, and how it could be deepened with the use of machines rather than diminishing it. With this, more job opportunities can be created with the use of machines rather than workers losing their chances with automation. And this comes down to five practical steps that one should take.

The first step to achieving the best could be in looking for ways to realize higher intellectual ground. People with advanced thinking and advanced abstraction level that outweigh the computers can always have opportunities to pursue. This means leaving things that are beneath humans to the machines while the people go for higher-order concerns instead.

Stepping aside approach involves using mental strengths that are out of the purely rational cognition level, and instead of going for the "multiple intelligences" aspect. This could be more of interpersonal or intrapersonal thing, including understanding how to work well with others and further understanding strengths, interests, and goals. Stepping in involves being able to understand how you can monitor and modify how computers work. Although the machine can take on the task of tasks, accountants can look for any mistakes that might be made by the automated programs and the users of the systems. Step narrowly approach is all about looking for a specialty in one's profession that can be economical if automated. This opportunity offers a great way to strike numerous deals that would otherwise be difficult to achieve. Since people design great machines, the step forward approach is all about creating the tools for the next generation of AI and computing. Today's inventions in this field may seem advanced, but there is still the need to invent for tomorrow, maybe for a different purpose than present automated aspects. The human need for better systems is almost insatiable, and this means more opportunities for invention, and jobs as well.

Employers need augmentation not only to ensure sufficient flow of job opportunities but also to overcome the rooted perception of automation mindset. However, employers should be convinced of the importance of the combination of computers and machines if augmentation is to succeed. The best thing to do is look beyond the perception that machines and people are substitutes and embrace the use of both on the same platform.

While globalization is nothing new, it is undeniable that this trend is rooted in several factors. It goes without saying that globalization comes with numerous benefits, but also packs a fair share of drawbacks as well. This aspect largely emanates from automation and connectivity, which has seen major revolutions in the tech world like the mobile internet. There is also a considerable gap developing between the winners and losers in the labor market, and these inequalities call for added attention. Such discrepancies are evident in industrial aspects, especially the manufacturing industry. The trend has remained unchanged for years. Skilled workers are getting the best of automation, while middle-class workers struggle with unchanging income and other effects of automation.

Quite often, workers' and skill displacement is all blamed on globalization, with the aspects of trade imbalance and lack of confidence in the ability of this factor to lead to economic growth. For many, the technological change poses more of a negative impact on the economy than a positive one. But the truth is we need to go further to understand the trend better. This can help in revealing where the industrialized societies are failing when it comes to adapting to technological changes that affect the workforce.

In essence, the society needs to adapt from working to embrace the aspect of creating the job, organizing and supervising it if we are to keep the economy moving and avoid inequalities in the society. If nothing is done fast enough things can only get worse with time. With all these changes happening, disruption of the business models as we know them might be inevitable and industries could as well suffer a similar fate. And this could go from manual workers and stretch all the way up to the most skilled personnel. Another thing that needs to be attended is the lack of adapting, which is seriously eating into the society's ability to make the most out of the opportunities that these technologies have made available. Such technological inputs can help boost education and health care with the progress that can help bring rise to further economic growth.

One of the best things to do is counter the negative effects of this disruption through increased investment in infrastructure and balance in trade agreements. Besides, enhancing the better environment for business creation can also go a long way to solving this problem once and for all. Above all, the inability to adapt to the changes in the technology as far as the workforce is concerned is among the things that need more attention. This can come easily if society can focus on adaption, retraining as well as redirection of the labor force. Another approach that will be beneficial is redesigning of education and career systems, which are at the center of the former industrial models. With these necessities at hand, humanity has to consider investing, which calls for collaborations in governments and educational institutions as well as a corporation in these sectors.

Conclusion

The realization that artificial intelligence is probably headed to taking over jobs and doing virtually everything is becoming a cause for panic among many people. However, the reality can be quite different; all it takes is the right measures to make the most out of this tech than letting it render millions of people jobless. Take, for instance, companies that are using artificial intelligence in most of the functions within, but still are creating entirely new opportunities for human labor. In these cases, the "influencers" are sourced from social media, especially people who have a significant number of followers. After analyzing the candidates, the company can then work with AI giants to bridge the influencer whose followers match the traits of the audience that the company in question is seeking.

Eventually, these brands can pay influencers to feature their products, which end up creating a completely new and reliable area of jobs. And this is not all, the use of AI is headed to inventions that are not existent today, as it has already done in the past. One of the new jobs this technology has created is the engine optimization position. All thanks to the internet, such unforeseen jobs and others have become a reality, and the chances are opening further along the way. As the advancement of AI carries on, the technology will help people learn to cope and prosper in this way of life. In this respect, AI classes are developed and taught to students to help them understand how it can be used for the benefits of everyone. This includes the technology helping you find any new jobs resulting from its use as well as understanding how to react to the ever-changing job markets, which will help reduce the rate of unemployment rather than worsening the situation, hopefully.

In the AI age, people will learn to focus on the jobs that utilize certain human strengths as creative thinking, social interaction, decision-making as well as the ones that require empathy, complex inputs, and questioning among others. After all, artificial intelligence can't think about data that is not present in the system, which is a capability bestowed in humans only. According to AI proponents, the technology will collaborate with humankind more than it is capable of becoming competition. For instance, the software can work by listening to the conversation in a conference room and searching the internet for related information content, and then it can provide the information on request. Proponents go further to state that AI can bring on board more information to people, which can be helpful in making better decisions, and taking human civilization to the next level.

Artificial intelligence is believed to be both good and bad to an extent in each respect. Whether it is a robot serving you at the gas station or AI technology taking away your job in the office, there will always be something to love about this tech as much as we might feel like running away from its engulfing grasp. But the bottom line is the fact that AI is not here to stop economic development, only to make things better for us, all that matters is the approach we take towards incorporating the tech into our lives and every aspect of our economy.

Automation is advancing in almost every aspect of today's world, and most people are afraid of this technology taking over their jobs. And despite the fact that changes seem inevitable in the most physical professions, knowledge-based jobs are believed to be on the line. However, the only question that remains crucial is whether automation can do more good than harm. But there is good news; this technology could be useful in transforming the business world for the better. Evaluating hard labor can offer an insight into just how essential the use of machines can become with time. Sometimes people are faced with errands they have to run either at home or anywhere else. But these duties could turn out to be backbreaking, and you might have to call someone to get the job done. However, the cost you will undergo in paying for the job is a set back on its own, and you will have to be there to oversee it.

So you end up spending as much time as you would in managing the results you wanted. You can avoid all this by using a machine and doing the task without having to wait. This shows that although elimination of jobs by these machines is taken negatively, there is a light to it in the long term. The reality, however, is that automation is bound to bring more prosperity in society in several ways. Highly educated workers can take on the job of developing these machines, while the less skilled counterparts use the machines in boosting productivity in the job. This strategy will lead to better output per any unit. The use of machines can bring a positive disruption of the economic status quo, which can be achieved by strategizing. Finding new ways of doing things can lead to success. So there needs to be a replacement of the production processes to bring value instead of managing costs. This is all about doing something unique and outsmarting your competitors.

The perception that you need to solve an existing problem to succeed might mislead you. Instead of looking for the problem to solve, take the unbeaten path and look for a non-existent problem. Stay clear of existing processes and products, where all you do is brainstorm on ways of improving on what has been done. Rather, it is better to create a new value that requires focusing less on what the others are doing and looking for something that you can do instead. These approaches are crucial for avoiding stepping right into the competitive environment and taking a different perspective that can help you beat the odds. And while focusing on this trend is crucial, you still need to make sure you are not throwing away money trying to minimize the costs. Bringing automation on board with the combination of these tactics will help you perform rather than leading to the loss of jobs. There is a need for preparation to take on any cases of workers displaced by artificial intelligence and robots taking over human labor. With this technology under fast development, the potential to disrupt livelihoods of many Americans is seemingly inevitable. And although this will not be a small deal, the country is in preparation to make sure any replaced workers do not suffer in the aftermath, which calls for better social safety measures.

According to a report by the U.S government, the use of aggressive policy action can help in fighting the condition that automation might leave in its wake. This means working in three prominent areas for the government to be ready for the wave of job displacement that will most likely be the result from the use of automation. One of the best ways is funding research aimed at learning AI and robotic fields in the U.S to keep the country's lead in the technology industry globally. The government can use this research to gain support for a wide range of workforce as well as focusing on fighting algorithm bias as far as artificial intelligence is concerned.

The other way is investing in as well as an increase in STEM education targeting the youth and retention of jobs for the adults in fields that have anything to do with technology. Computer science for students and expansion of workforce nationwide can also be used through investing to help workers in America stay competitive in the worldwide economy. Modernization and strengthening of social networks at the federal level through means like public health care, welfare, and food stamps as well as unemployment insurance can also go a long way. Besides, the report advocates for an increase in the amount of minimum wage, stronger unions, and payments for overtime as well as enhancing the bargaining power of workers. However, the current government's policies are not consistent with these recommendations, with some discrepancies in areas like public education. It is worth noting that this approach significantly touches on a better understanding of computing among other AI-related aspects. The government will need to do more than it is already doing if it wants to achieve the recommendations laid out in the report. This can be crucial for achieving better understanding on how to counter human labor replacement by machines and the effect this trend might have on the country in the end.

As recommended in the report, policies will help in facing the complexities of social services. This approach will be essential in protecting millions of Americans from imminent displacement as robotics and automation, as well as artificial intelligence. Besides, the government needs to monitor the support on the competition in the artificial intelligence field. This comes into play when you consider the fact that companies that have the most data are capable of creating the best products. Eventually, such a trend can inhibit startups from gaining a much-needed competitive edge, and as it turns out, such starter enterprises are the backbone to creating more jobs that can cover the gap resulting from robot replacement.

Artificial intelligence has been with us for years, from the Egyptians' automatic water clocks to today's robots. The contemporary technology in industrial revolution extends as far back as the 1800s. But the trend is headed for the next level with automation affecting the current processes and its programmable logic changing almost everything. One of the things that this technology is good at is the opportunity that tech has created with the chance of achieving greater results. Such tasks can be impossible, dangerous or outright difficult for humans. Besides, robotics has offered an opportunity to automate complicated tasks just like humans would do. These ideas can help to identify a transaction with the risk of fraudulence especially on a customer's patterns of credit card spending.

Moreover, sensors in objects we use on a daily basis attribute to offering positive results of machines. Lights in air conditioners, televisions, and other tools are among the best sensors to get in touch with your appliance. Not to mention, these sensors are great when it comes to tracking speeding cars as well as helping identify how many spaces are available in a vast parking lot.

Driverless cars, chatbots, and 3D printing machines are among the other aspects where robotics is becoming essential by the day. This tech are not only making life easier for humans but also enhancing productivity and most other aspects as well. The hard reality, however, is the fact that as much as automation can affect our lives for the better, it can also turn out to be a challenge altogether. One of the setbacks that this technology can present to humanity is the scarcity of jobs. If there is a massive shift towards adopting autonomous cars, for instance, then the loss of jobs for most drivers might be inevitable. Nevertheless, this will most likely come with a low-cost opportunity for starter investors to get into the business and get a livelihood out of it. Not forgetting, software engineers can also find an opportunity for new jobs, especially when it comes to programming and data analytics for these cars. Another aspect faced with uncertainty is the customer service representatives who risk losing their jobs to chatbots. But this could also turn out to be a great chance for service delivery customers and people practicing data science analytics. This is because tech will lead to the need for complex programming as well as artificial intelligence, which can similarly create job opportunities on one end despite overriding others on the opposite side. In the history of industrial revolution, there has been the creation of jobs and replacement of others. And as technology goes to the next level, the same will most likely recur. It is safe to expect automation to come with benefits and downsides to it as well. It looks like robotics are not out to push humanity to the edge, but most likely create as many opportunities as it takes away. The best thing to do is brace for change and be prepared to take it on.

References

Bennett, S. (1993). A History of Control Engineering 1930-1955. London: Peter Peregrinus Ltd. On behalf of the Institution of Electrical Engineers. ISBN 0-86341-280-7.

Dunlop, John T. (ed.) (1962), Automation and Technological Change: Report of the Twenty-first American Assembly, Englewood Cliffs, NJ, USA: Prentice-Hall. * Dunlop, John T. (ed.) (1962), Automation and Technological Change: Report of the Twenty-first American Assembly, Englewood Cliffs, NJ, USA: Prentice-Hall.

Ouellette, Robert (1983), Automation Impacts on Industry, Ann Arbor, MI, USA: Ann Arbor Science Publishers, ISBN 978-0-250-40609-8.

Trevathan, Vernon L. (ed.) (2006), A Guide to the Automation Body of Knowledge (2nd ed.), Research Triangle Park, NC, USA: International Society of Automation, ISBN 978-1-55617-984-6, archived from the original on 2008-07-04.

Frohm, Jorgen (2008), Levels of Automation in Production Systems, Chalmers University of Technology,

www.ingramcontent.com/pod-product-compliance
Lightning Source LLC
Chambersburg PA
CBHW071726170526
45165CB00005B/2175